YOUR KNOWLEDGE HAS VALUE

- We will publish your bachelor's and
  master's thesis, essays and papers

- Your own eBook and book -
  sold worldwide in all relevant shops

- Earn money with each sale

Upload your text at www.GRIN.com
and publish for free

Ram Krishna Pandey, Abhinav Mishra, Rupesh Tomer

Fruit Trade in India. Matlab Programming for Prediction of monthly Arrival and Prices of various Fruits

GRIN Publishing

Bibliographic information published by the German National Library:

The German National Library lists this publication in the National Bibliography; detailed bibliographic data are available on the Internet at http://dnb.dnb.de .

Imprint:

Copyright © 2009 GRIN Verlag GmbH
Print and binding: Books on Demand GmbH, Norderstedt Germany
ISBN: 978-3-656-93904-7

This book at GRIN:

http://www.grin.com/en/e-book/295348/fruit-trade-in-india-matlab-programming-for-prediction-of-monthly-arrival

Fruit Trade in India. Matlab Programming for Prediction of monthly Arrival and Prices

RAM KRISHNA PANDEY

ABHINAV MISHRA

RUPESH TOMER

POST HARVEST PROCESS AND FOOD ENGINEERING DEPARTMENT, COLLEGE OF TECHNOLOGY,

GOVIND BALLABH PANT UNIVEERSITY OF AGRICULTURE AND TECHNOLOGY

PANTNAGAR -263145 (U.S.NAGAR), UTTARAKHAND INDIA

1. INTRODUCTION

The fruit trade in India is conducted through a network of wholesale markets called Mandies. These Mandies represent a complex interaction amongst seller (grower), buyer (trader) and the regulatory agency (government). Therefore, for a properly managed post-production system fruit Mandies are the nodal points through which all the primary production of the concerned production catchment must pass. Thus a suitably designed modern Mandi could provide a unique opportunity for processing of fruits, which may involve receiving, cleaning and grading, drying, storage and disposal. Such a system is also expected to afford substantial reduction in primary losses of the total fruit production. However, for a rational and logical design of the Mandi system, including the various processing facilities, the characteristics of inflow i.e. fruit arrival pattern must be known, or be capable of being forecasted.

These arrival patterns are strongly time variant. One of the widely applied techniques to describe the time dependent stochastic processes is time series analysis. Since, the Mandi arrivals and prices form an ordered sequence of observations; they also represent a time series. Therefore, their time dependence could also be established using time series analysis techniques. Once the mathematical model to describe the time dependent structure of the series is developed it could be extended to generate future forecasts and help in Mandi design. The present study is a step in this direction. Specific objectives of the study are:

1. To analyse fruit arrival and price patterns using the historical arrival and price records of Haldwani Mandi,
2. To develop a program in MATLAB 7.0 for time series modelling in order to describe the time dependent structure of the fruit arrival and price series; and
3. To extend the time series models for prediction of monthly arrival and average monthly price value of the series.

2. REVIEW OF LITERATURE

Generally, there are two approaches to model the forecasts. One is the so-called classical econometric approach which involves an analysis of the factors on which the variable to be predicted depends. It concentrates on the specification of a system of behavioural equations which characterize the underlying variable to be predicted by means of multiple regression or simultaneous equation techniques. Another category of forecasting models is the time series approach, which is in contrast to the econometric approach in that no independent variables are used. This approach bases its predictions solely on current and past behaviour of the variable, by assuming that the actual observed series may be considered as a realization of some stochastic process with a structure that can be characterized and described. Examples of this approach include autoregressive (AR) models, moving average (MA) models, exponential smoothing (ES) models, a combination of AR and MA models and the Box-Jenkins (BJ) techniques.

Baudentistl (1971) examined marketing trends for the total milk supply and individual dairy products in Austria for 1960-70 period with a view to predict likely future trends in the market from developments in the past. Total milk supply during the period increased steadily until 1967. After a stagnation period, supply increased again from 1969, although to a lesser extent.

Singh and Sidhu (1972) investigated the trends of market arrivals and prices of wheat, paddy, maize and gram in different size markets of the Punjab state, for the period from 1964-65 to 1969-70. The arrivals of all the four commodities were much higher in the bigger markets because of high prices prevailing there. Arrivals of wheat, maize and gram in large markets were between 60 and 70% and those of paddy were about 85% of the total arrivals. The major portions of the arrivals were during the post-harvest period. The arrivals varied inversely with the prices.

Blyn (1973) investigated into the price series correlations as a convenient measure of market integration. Eight year collection of monthly wheat price data in eight Punjab markets and Delhi was analysed to eliminate trend and seasonal influences and detrained series were then correlated. He concluded that correlation among prices ruling in a set of markets was imperfect measure the degree of market integration and that correlation among the prices adjusted for trend and seasonality was a better measure.

Singh (1975) examined arrivals and prices for 1960-61 to 1970-71 periods in the Khanna market of Punjab for the possible correlation. The results showed high correlation. The results showed high correlation between arrivals and prices i.e. when the arrivals were high, prices were low and vice-versa. This is in agreement with the findings of Singh and Sidhu (1972) and Venkataramanan and Murlidharan (1972). He also observed that the lagged prices of groundnut affected its production.

Hanssens (1977) illustrated the use of invariable and multiple time series analysis in the development of descriptive marketing models. He used Box-Jenkins ARIMA model for modelling the individual time series. His approach is different earlier investigators, in that none of them employed time series analysis for examining market arrival patterns and its correlation with the market price. He also explored the relationship between econometric models and time series models and argued that a combined use of both techniques is most favourable for marketing model building.

Singh and Sindhu (1979) examined seasonal variations in production, marketed surplus and process of milk in the urban and rural areas of Punjab in 1977. They found that the production in urban areas varied from 85,111 litres in May to 98,666 litres in November and from 28,750 litres in June to 61,250 litres in January in the rural areas. Marketed surplus of milk accounted for 91.5 % and 71.5% of total milk production in the urban and rural areas respectively. The marketed surplus varied with the level of production, consumption and price over the different months.

Sharma and Roy (1979) analysed the trends in Indian food grain consumption and factors influencing it. They observed that per capita consumption and factors influencing it. They observed that per capita consumption declined due to rise in food prices. The other factors which might have caused this decline are identified to be the shift in taste away from food grains and deterioration and distribution of income.

Jain and Kaul (1980) analyzed the seasonal fluctuations in the potato market and the existence of cycles and their periodicities using series data relating to arrivals and prices from four major potato markets in Punjab. A multiplicative time series model was fitted to the data to analyzed trend, seasonal and cyclic variations in potato prices and arrivals. Prices showed an upward trend but large fluctuations in a year confirmed by cyclic variations with three year cycle period. No attempt is however, made for forecasting future arrivals and prices.

Boyle (1980) made a time series analysis of monthly pig deliveries at Irish bacon factories using Box-Jenkins technique. The forecasts were made for 24 months ahead with an average monthly error of 4.4 %.

Luandahl and Peterson (1982) calculated monthly correlation coefficients, using 1969-74 price data for five food grain products from Haiti markets, to show seasonal pattern and what they indicate about the structure of the marketing network. The numbers of markets considered for each product were: 10 for rice, 8 for millet, 20 for maize, 11 for ground maize and 15 for red beans. The results showed a lower correlation during the harvest months when most of the deliveries were made. The patterns show by the correlation coefficients for all the products were contrary to the expectation, probably because of the uni-directional trend.

Ngenge (1982) used time series models to investigate the dynamic relationship between the weekly cash prices of corn, sorghum and soybeans in three regions of the USA. In general, multivariate models performed better than univariate models, all markets were inefficient and most markets adjusted to changes in supply-demand conditions within a week, although residual adjustments took several weeks.

Singh and Gupta (1982) analysed grain arrival patterns and marketing practices of five food grains (paddy, wheat, maize, rapeseed and gram) for Rudrapur grain market of Uttar Pradesh. The maximum weekly arrivals varied from 3845 T of wheat to 180 T of gram. Annual volume of arrivals for wheat and paddy over the period 1971-73 were comparable with an average of 16470 T followed by maize (2400T), rapeseed (1940 T) and gram (660T). The combined monthly volume of all grains varied considerably (10 - 12280 T) over the three year period. No effort was made to analyse the trends and seasonality and the forecasting models which is an important aspect of effective planning of Mandi operations and modernization programmes.

Narain (1984) analysed the grain arrival and price patterns in a wholesale grain market (Mandi) for trend, seasonality and periodicity in order to develop forecasting models for the grain arrival process so as to rationalize an important input to grain Mandi system design. Historical time series data on monthly arrivals and average monthly prices was collected from the Rudrapur Mandi records for the period 1968-69 to 1983-84. Arrivals and prices of five grains (wheat, common paddy, Basmati, rapeseed and masoor) were considered. These taken together constituted more than 90 % of the total Mandi arrivals and represented the three major classes of food grains i.e. cereals, oilseeds and pulses. Forecasting models were developed on the basis of first 168 months (July 1968 to June 1982) data using time series analysis technique. Forecasts were generated for the next 36 months (July 1982 to June 1985). These forecasts were compared with the actual arrivals for July 1982 to June 1984 period. Model adequacy was tested by the independence of the generated residuals.

Yuan (1985) developed a model of fruit and vegetable wholesale markets in Taiwan, based on size, location, transport costs, flows and quantity and destination. The results of the model indicated that 27 fruit and vegetable markets were needed in southern Taiwan by 1981, and a further six by 2001, i.e. one per region. Size should vary according to volume of trade. Transport costs should be considerably reduced by increasing the number of markets. The township is recognized as the base unit for market planning.

Kallolo *et at.* **(1988)** studied the relationship between grade and quality, and between price and quality of a product using groundnuts as an example. Data were collected from three regulated markets (Gokak, Hubli and Dharwad) in Karnataka state, India. A grade response model was fitted for each of the selected markets with grade value of the sample as dependent variable. To test these models multiple linear regression analysis and step-wise linear regression analysis were used. The study revealed that the grade value and price are closely related. The main variables affecting price are, dryness of the pods, maturity, soil content of the pods, shelling percentage, oil percentage and moisture percentage. The existing grading systems in the regulated markets were far from satisfactory.

Wegner (1989) in his theoretical and empirical study analysed production structure and developed, demand, international trade and agreements and prices for selected fruits. These include three with wide spread productions (apples, pears, kiwifruits), three tropical fruits (bananas, pineapples, mangoes) and three subtropical fruits (sweet citrus fruits, grapefruit, avocados). Conclusion are that although demand for all type of fruit is increasing with rising income (detailed demand analyses relate to the USA and GFR), there is a tendency for production of the more popular varieties to expand more rapidly than demand, with consequent choking of the market. Changing consumer tastes as well as rising incomes are important determinants of demand and are increasing the market for different varieties of fruits and particularly for exotic fruits. This helped by the increasing liberalization of trade products which greatly benefiters' developing countries.

Pouch (1997) studied that in year 1994 world fruit production reached 388 Mt, of which 10 % were exports of fresh fruit. In the same year production of temperature climate fruit (apples, pears, peaches, plums, apricots, grapes and kiwis) reached 125 Mt, of which 8 Mt were exports, a 30 % increase since 1985. There has been a considerable increase in demand for fresh fruit has attracted many producers in the southern hemisphere, particularly Chile, Argentina, Mexico and Brazil. The general decrease in prices for tropical product exports has resulted in a diversification of production towards temperate climate fruit, which are in demand in the northern of hemisphere during the off season. A more detailed analysis of world apple, table grape and kiwi markets is presented and the position of France is examined briefly.

Yadav *et al.* **(1999)** studied the probability analysis of some of the fruits and vegetables arrival in Haldwani Mandi for eight years and ten years data, using the same five distributions and concluded that all the five distributions were good at five percent level of significance with probability factor highly significant. It was seen that for one month arrival at 80 percent probability level percentage deviation was minimum for Normal distribution and at 90 percent probability level it was minimum for Log Normal distribution and at 50 percent probability level percentage deviation was minimum for Log Pearson type HI distribution. The estimated values of these distributions and the actual arrival in next two years were compared. For two months consecutive maximum annual arrival, percentage deviation was minimum for normal distribution at 80 percent probability level and for Log Normal at 90 percent probability level and at 50 percent probability level percentage deviation was minimum for Log Pearson type III distribution. Picchi (2002) studied the factors affecting the interval between picking various fruits and their appearance on supermarket shelves is discussed. The categories considered are; most perishable, moderately perishable (apples, pears, citrus, nuts, etc.) and frozen or dried fruits. Times taken in transport, loading and unloading, sorting, storing, packaging and labelling clearly vary with the type of produce. Management of the stock on arrival in the supermarket is also considered. The topic is illustrated with the help of flow charts and diagrams.

Tijskens *et al.* **(2003)** analysed that a considerable proportion of the fresh produce (such as fruit, vegetables and flowers) sold around the world is exported. While most of the export is successful, a disturbingly high proportion of shipments encounter quality problems on arrival at destination. New technologies have been developed for monitoring refrigerated shipping containers that makes it possible for actual storage conditions during transport to be monitored in real time. This enables speedy rectification of any problems and alerting to any quality problems on arrival. Besides the considerable benefits that this technology achieves for all involved in the shipping and fresh produce industry, it is possible to interface storage prediction software and real time monitoring of storage conditions to provide a premium service to the industry that can fine tune the quality delivered to market requirements. Advantages of such a model are that decisions can be made before shipment as to potential storage life remaining for a particular crop (e.g. apples stored in controlled atmosphere prior to export). Then, if there is a problem for a particular voyage length, the shipment can be cancelled or storage conditions adjusted to assure a longer storage life. The development of a fully robust model to cover a wide range of products, a wide range of pre-shipment conditions and a range of problems during shipping will obviously take some time but the benefits will be considerable. Fitting a wide range of models to data for over 30 crops suggests that the best choices are the exponential and a modified quadratic model.

Anand and Sareen (2007) worked on probability analysis of fruits and vegetable arrival in Haldwani Mandi. They used the past years' data from the Mandi records and developed a model to describe the probability of arrival in each month.

Singh and Sajwan (2008) used ratio to trend method and simple average method to analyse seasonality in the time series of monthly arrival and prices of selected fruits at Haldwani Mandi. They also analysed the pattern of the arrival and average monthly prices for the period 1990 to 2007. However, their work did not include the extension of the time series in order to predict the future values of arrival and prices of these commodities at the Mandi.

Pandey and Mishra (2009) worked on the application of artificial neural networks to analyse and predict monthly arrival and average monthly prices of wheat and paddy at Rudrapur Mandi. They used various weather parameters in order to understand the pattern of the time series. However, the drawback of their work was that the catchment area of the Mandi was very large and hence the weather parameters that they took (of Pantnagar) did not represent the whole catchment area.

3. MATERIALS AND METHODS

The present study aims at modelling and forecasting fruit arrivals and prices in a wholesale fruit market which, in India, is popularly called Mandi. The state of Uttarakhand has 19 Mandies. Haldwani Mandi, which is located in the vicinity of Govind Ballabh Pant University of Agriculture and Technology, is selected for collecting historical time series data on fruit arrivals and prices. A brief discussion of the Mandi, its operations, methodology of data collection and analysis and development of forecasting model follows.

3. 1 Haldwani Mandi and its Operations

The data was collected from the office of Uttarakhand Krishi Utpadan Mandi Parishad - HALDWANI. Earlier it was known as Haldwani Mandi which came into regulation on SEP

1-1971 but the regulation of UKUMP was done on DEC 12- 2000. The total area is 43.59 acres and the main Mandi areas include Haldwani and Kathgodam. The sub-Mandi areas include Mukhani, Lamachaur, Kaladhungi, Lalkuan, Bhawali. The main objectives of Mandi are-

1. To eradicate the extraction of unauthorized taxes and to lessen the various commercial expenses. To provide competitive rates to the farmer for his production.
2. To eradicate the problem in weighing and doing regular check out of weighing machines and other instruments.
3. To establish proper Mandi Samities, this will favour farmers and can do their welfare.
4. To make pricelist available to farmers and traders.
5. To construct mandi 'sthals' and equip them with proper facilities.
6. To contruct best warehouses and stores.

The regulated products of Mandi are given in Table 3.1

Table 3.1 Regulated products of Uttarakhand Krishi Utpadan Mandi Parishad - HALDWANI

Food grains	Fruits	Vegetables	Pulses	Others
Paddy	Apple	Potato	Urd	Ghee
Rice	Banana	Ginger	Moong	Fish
Wheat	Mango	Radish	Arhar	Wood
Maize	Orange	Cauliflower/Cabbage	Lobhia (seed)	Gum
Bajra	Peach	Tomato	Masoor	Honey
Juar	Litchi	Onion		Cotton
Jau	Grape	Ladyfinger		Jute
Bejhar	Keenu	Carrot		Sakkar

Jaie	Chicu	Lemon		Gur
Gram	Mosmbi	Tinda		Egg
Soyabean	Parwal	Brinjal		Groundnut
Dancha(seed)	Guava	Garlic		Til
Atta	Grapefruit	Guar		Mahuwa
Maida	Ber	Sanie (seed)		Linseed
Suji	Cocunut	Chilly (peak)		Castor
Paddy (seed)	Jackfruit (reap)	Pea		Sunflower
	Peaches	Green pea		Sanie (fiber)
	Loquat	Parwal		Tobacco
	Aonla	Cucumber		Sakkar
	Pumelo	Pumpkin		Khandsari
		Whitegrid		Mahuwa
		Jackfruit (green)		Amchur
		Sweet Potato		Turmeric
		Bhusha		Ramdana
		Mushroom		Jeera
		Chilly (green)		Sauth
				Cashewnut

3.2 Data Collection

Since Haldwani is located just at the feet of the Kumaon region, fruits constitute a large fraction of arrival at the Mandi. For the purpose of present study, the arrival of five fruits- Apple, Banana, Mango, Orange and Peach have been taken and prices of Apple and Banana have been taken. The monthly prices of other fruits could not be analysed because of the uncertainty in the prices of these fruits in the months when there was no arrival of these fruits.

Historical data on monthly fruit arrivals and average monthly prices for the past 19 years (January 1990 to April 2009) in respect of the five fruits considered are collected from the Mandi Samiti records. The time period corresponds to the calendar year i.e. January to December. Historical data on fruit arrivals and monthly average prices for different fruits are given in Table A.1 to Table A.5 and Table B.1 to Table B.2. Monthly fruit arrivals are recorded in quintals and average monthly fruit prices are recorded in Rupees per quintal. Out of the 19 years historical data, 17 years data i.e. data for the first 204 months (January 1990 to December 2006) is used for developing and fitting the forecasting model while the remaining 28 months data has been used to test the accuracy of the forecasts generated.

3.3 Data Analysis

Data analysis includes arrival and price patterns, trend analysis, seasonal variations and computation of seasonal indices. Details of the procedure followed in the analysis are given below.

For the purpose of examining the pattern of fruit arrivals and their prices, each year has been divided into 12 months. The patterns are shown in the figures 3.1 to 3.7.

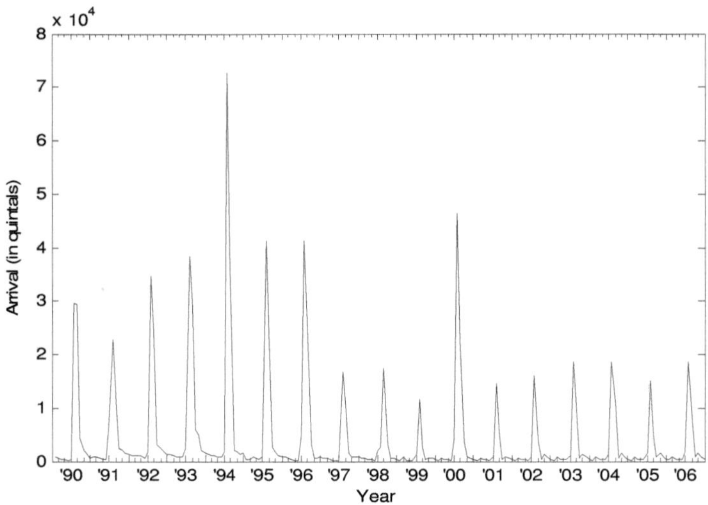

Fig 3.1 Monthly apple arrivals

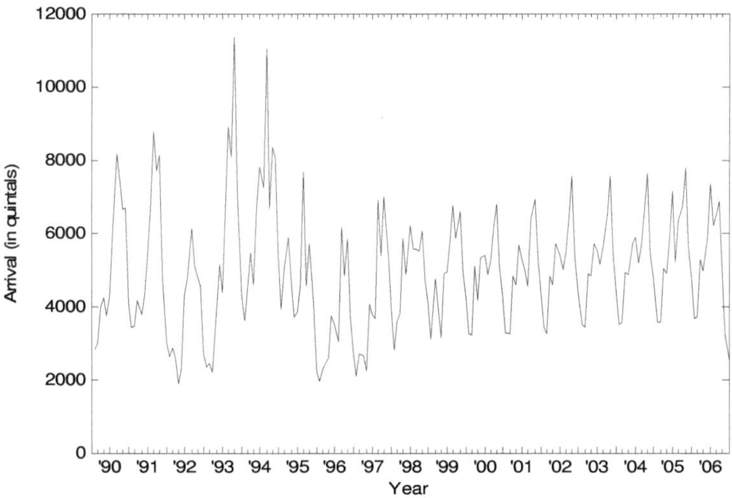

Fig 3.2 Monthly banana arrivals

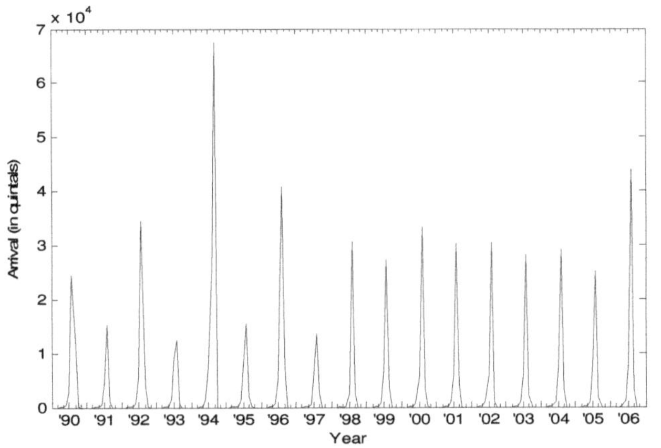

Fig 3.3 Monthly mango arrivals

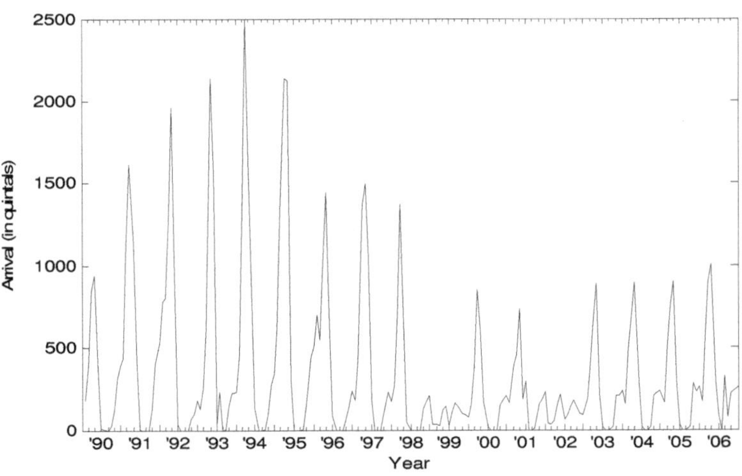

Fig 3.4 Monthly orange arrivals

14

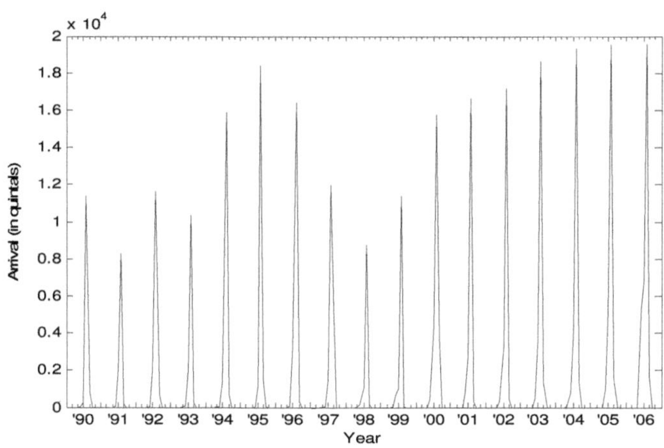

Fig 3.5 Monthly peach arrivals

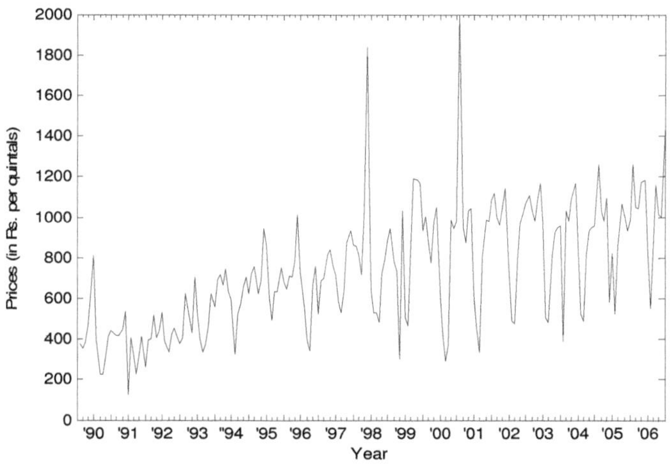

Fig. 3.6 Monthly apple prices

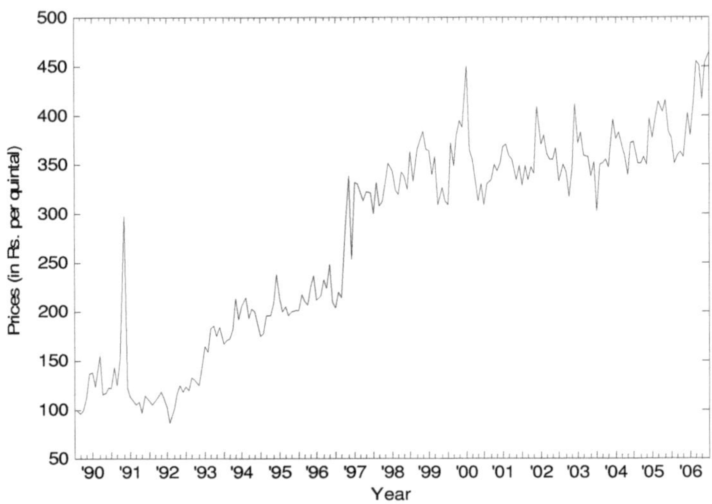

Fig3.7 Monthly banana price

3.4 Programming for computation of trend

For analysis of seasonal variations, the Least Square Method was employed. A program in MATLAB 7.0 was developed using the following algorithm.

Step1: Enter the data of arrival/ prices corresponding to the selected fruit month wise for the years under analysis.

Step 2: Calculate the average value for each month over the years selected for analysis.

Step 3: Now, calculate the average value of monthly arrivals/ prices for each year under analysis. The values corresponding to each year will serve as average per month (y value) for each year.

Step 4: Now assign the value of coded x against each year as mentioned under the description of Least Square Method.

Step 5: At this stage, we are able to obtain every parameter used to calculate the value of 'a' and 'b' described in the Least Square Method.

Step 6: Now we have the fitted equation $Y = a + bX$ for the trend value of each year, where Y stands for arrival/ price and X corresponds to the year's coded value.

Step 7: The values obtained by the above method for each year corresponds to the middle of the year, i.e., 1st July. Therefore, adjustments have to be made for each month by taking the average of the difference between the trend values of two years over the twelve months. After the adjustments have been made for each month, we obtain the trend values for each month.

Based on these steps a flow chart was developed (shown in Fig. 3.8)

3.5 Programming for computation of seasonal variations

For analysis of seasonal variations, the Ratio to Trend Method was employed. A program in MATLAB 7.0 was developed using the following algorithm.

Step 1: Enter the monthly trend values of arrivals/ prices corresponding to the years under the analysis.

Step 2: Calculate the medians of these trend values for each month over the years under analysis. The median is taken so that the effect of extreme values is eliminated.

Step3: Find the sum of these medians. If it is not equal to 1200, then adjust these values so that their sum is equal to 1200. This can be done by multiplying each value by 1200 and then dividing by the sum obtained by adding the medians.

Now, these values are the monthly seasonal indices. A flow chart was developed using the steps involved in the algorithm of ratio to trend method (Fig 3.9)

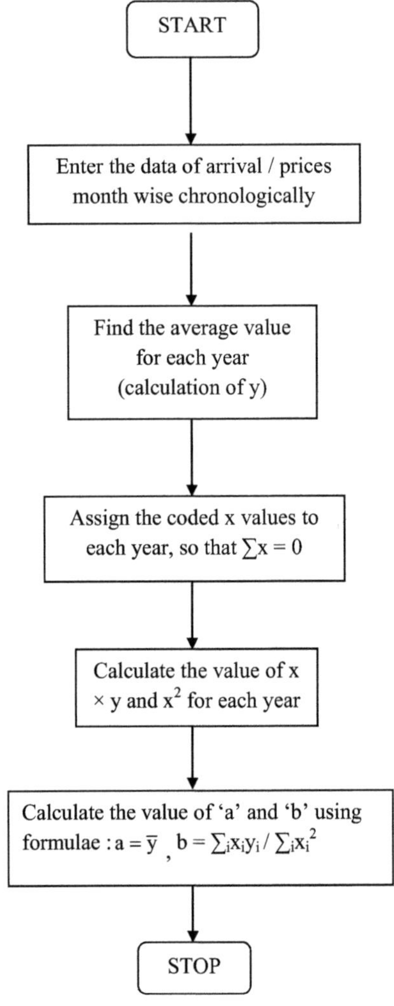

Fig 3.8 A flow chart representing the algorithm of least square method

Fig 3.9 A flow chart representing the algorithm of ratio to trend method

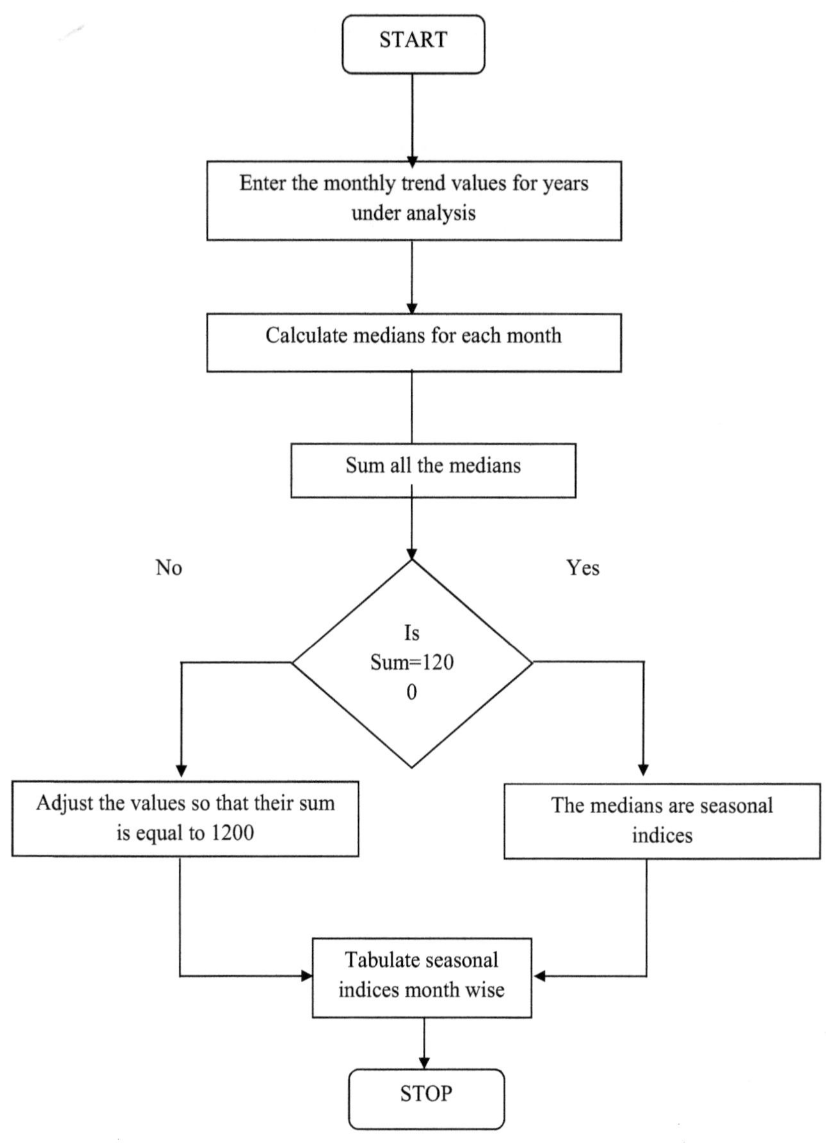

4. RESULTS AND DISCUSSION

The objective of this study is to analyse fruit arrival and fruit price patterns in a fruit Mandi and to formulate an appropriate mathematical model to describe the time dependent structure of monthly fruit arrivals and average monthly fruit price so as to make futuristic forecasts of the arrivals and prices. These forecasts would form an important input for rational design of the Mandi system. For the purpose, data series for monthly fruit arrivals and average monthly prices for five fruits (mango, banana, apple, peach and orange) are analysed. The data series both for fruit arrivals and prices consisted of the monthly data for January 1990 to April 2009. The first 204 months data is used to analyse the pattern and to construct the various models, while the last 28 months observations are used in testing the accuracy of forecast models.

4.1 Trend Analysis

For trend analysis, Least Square Method was adopted. A computer program using MATLAB 7.0 was developed for this analysis. The objective of this program was to obtain a fitting equation $Y_c = a + b\,X$, where 'Y_c' indicates the trend value, 'a' and 'b' are constants to be calculated and 'X' denotes the coded value for the year as discussed in the Least Square Method. 'a' is the intercept at the y- axis whereas 'b' indicates the slope of the line. Higher the value of 'b', higher will be the rate of increase in the arrival/ prices with the passage of time.

For the arrival of apple, the values of 'a' and 'b' are 4407 and -306 respectively, which clearly indicates that there has been a gradual decrease in the arrival of apple at the Mandi. The reason may be establishment of new Mandies in the area or establishment of new industries which consume apple. In the case of its prices, these values are obtained as 730 and 38 respectively, implying that these prices are increasing strongly.

For banana arrival, the values of 'a' and 'b' are 4970 and 19 respectively, showing a slight increase in the arrival of this commodity in the Mandi. For the monthly average prices of banana; these values come out to be 286 and 20 respectively, which also indicates an increasing trend of the average monthly average prices of it.

In the case of mango arrival these values are 3449 and 11 respectively, meaning that there is a gradual increase in its arrival. For orange arrival, these values as obtained by the program are 384 and -15 meaning that there has been a decrease in its arrival, and for peach arrival these values are 136 and 0.5 respectively, showing that its arrival is increasing, but at a very lower rate.

These values are further useful in calculation of seasonal indices.

4.2 Seasonal variations in arrivals and prices

The indices for monthly arrivals and prices are calculated using Ratio to Trend Method. A computer program using MATLAB 7.0 was developed for this analysis. The analysis of seasonal variations shows the effect of seasonality on arrivals and prices. The seasonal indices for fruit arrivals and prices are given in table C.1 and C.2 respectively. The seasonal indices for each month have been plotted and are shown in figure 4.1 to 4.5.

It is clearly evident from table C.1 that in apple, the arrivals are maximum in July with an index of 626.65 and minimum in May with an index of 5.65. By closely observing the seasonal index graph of Apple (Fig. 4.1), we find that the arrival of apple is increasing from January to February and then there is a decrease up to the month of May, after which it increases drastically up to July and then it decreases till December. The prices behave inversely for which the index is lowest in July (60.74) and highest in May (127.48).

In banana, the arrivals are maximum in October with an index of 136.30 and minimum in January with an index of 65.02. The prices seem to be affected by the arrival but do not behave completely inversely and the index is lowest in October (90.44) and highest in July (105.90). One of the major reason for which the values of prices remain irrespective of the arrival to some extent seems to be that unlike apples, the seasonal indices of the prices of banana have relatively lower range (the maximum being 105.90 and the minimum is 90.44), i.e., the prices do not vary too much throughout the year.

In the case of mango, it is quite evident from the table 4.1 that the probability of its arrival is only from March to August, since the seasonal indices for the months from September to February are zero for the period under analysis. The arrivals are maximum in July with a seasonal index of 876.06 and from September to February, there is no arrival. However in March and April also, there is very little arrival of mangoes at the Mandi.

The arrival of orange is maximum in the month of April with a seasonal index of 371.17, whereas it is completely absent in the Mandi in the Months of July and August.

However, in the months of June and September also, there is very little arrival of orange with seasonal indices of 7.01 and 0.24 respectively.

In the case of peach, the arrival is observed only in four months, i.e. from May to August, the highest being in the month of May with a seasonal index of 439.89. In the rest of the eight months, there is no arrival of peach.

4.3 Future Forecasts:

Based on the study of trends and seasonal variations, the monthly arrival and average monthly prices were forecasted for the time period of 36 months (January 2007 to December 2009). For the validation of forecasting, the actual values of first 28 months (January 2007 to April 2009) corresponding to this forecasting were used. A comparison between these actual and forecasted values has been shown in Fig. 4.6 to Fig. 4.12.

In the case of monthly arrival of Apple, the peak arrivals estimated in the years 2007, 2008 and 2009 were10579, 8654 and 6729 respectively (shown in Fig.4.6), whereas actual values of peak arrival in 2007 and 2008 were observed as13342 and 11896 respectively, which shows that there has been a maximum error of 26.89% in forecasting the peak arrival. These peak arrivals were observed in the month of July.

In Banana arrival, the forecasted peak arrivals were 7013, 7039 and 7066 for the years 2007, 2008 and 2009 respectively (Fig. 4.7). The actual values corresponding to these peaks for the first two years were 6489 and 6879 (all in the month of October), which gives a maximum error of 8.07%.

For the arrival of Mango, the peaks were observed in the month of July. The values corresponding to these peaks for the years 2007, 2008 and 2009 were predicted as 31113 Q, 31218 Q and 31323 Q respectively, whereas the actual values for these peaks as collected from the Mandi for the first two years was 32005 Q and 32787 Q (Fig. 4.8), giving a maximum error of 4.78 %.

In the case of monthly arrival of Orange, the peaks were forecasted as 923 Q, 865 and 807 for the years 2007, 2008 and 2009 (in the month of April), whereas the actual values corresponding to these years were 876 Q, 856 Q and 823 Q (Fig. 4.9), resulting in a maximum error of 5.36 %.

The peak arrival of Peach was forecasted as 621 Q, 623 Q and 626 Q for the years 2007, 2008 and 2009 respectively. The actual values corresponding to these peaks in the month of May were observed as 573 Q and 572 Q for the years 2007 and 2008 respectively (Fig. 4.10). This reflects a maximum error of 8.91%.

In the case of the average monthly prices of Apple, the peak prices were observed in the month of May. The estimated peaks for the years 2007, 2008 and 2009 were 1369, 1419 and 1470 (all in Rs. per Q), whereas the actual values for these peaks for the years 2007 and 2008 were 1678 Rs. per Q and 1647 Rs. per Q respectively (Fig. 4.11). This resulted in a maximum error of 18.41%.

The average monthly prices of Banana were very irregular in nature. However, on the basis of the trend and seasonal analysis, the forecasts were made for these and the peaks were forecasted for the three years 2007, 2008 and 2009 as 496 Rs. per Q, 517 Rs. per Q and 539 Rs. per Q (in July). The true values corresponding to these peaks were 458 Rs. per Q and 566 Rs. per Q (Fig. 4.12). Thus the maximum error corresponding to peak values was 12.88%.

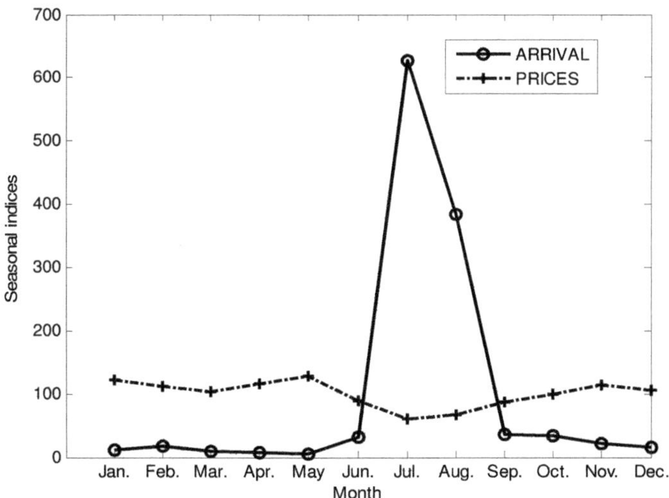

Fig 4.1 Seasonal indices for apple

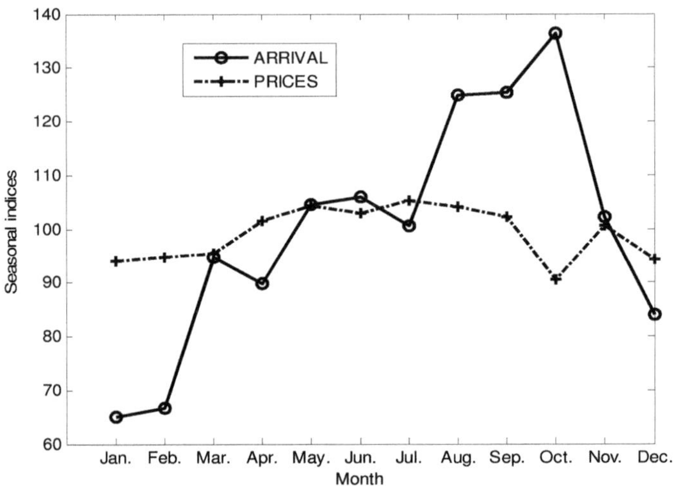

Fig 4.2 Seasonal indices for banana

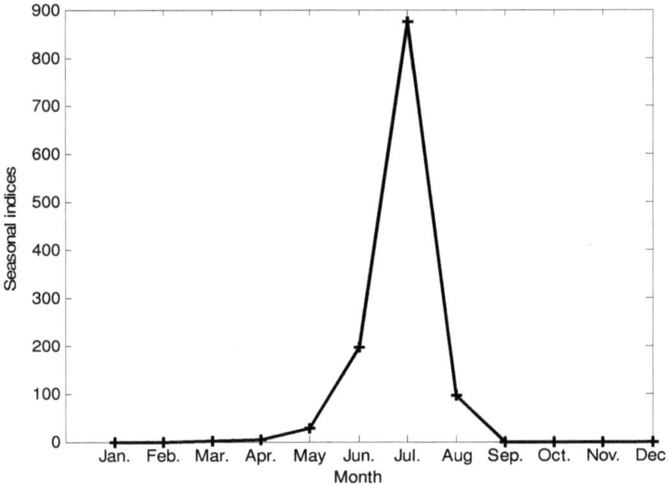

Fig 4.3 Seasonal indices for mango

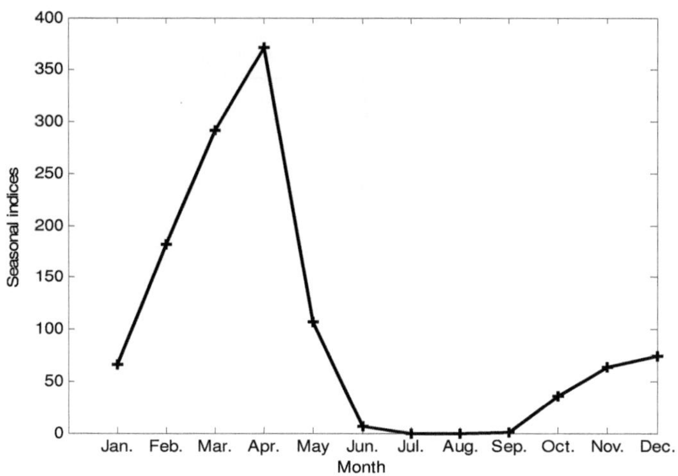

Fig 4.4 Seasonal indices for orange

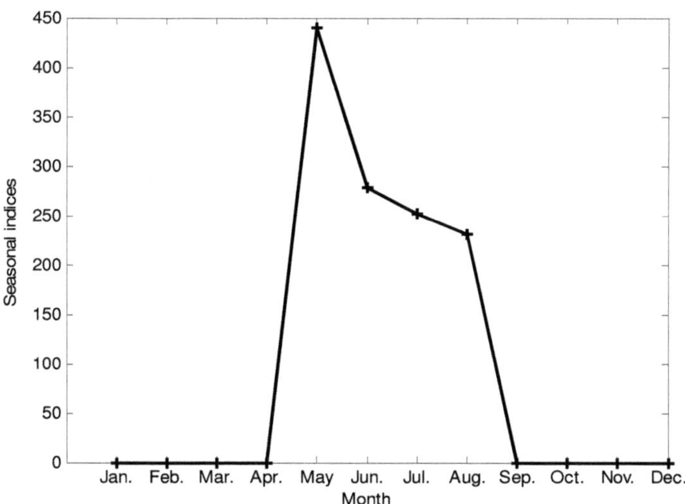

Fig 4.5 Seasonal indices for peach

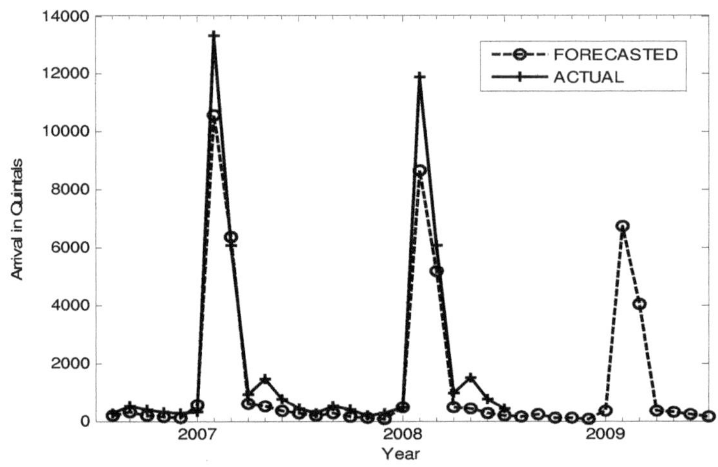

Fig 4.6 Graphical representation of actual and forecasted arrivals of apple from January 2007 onwards.

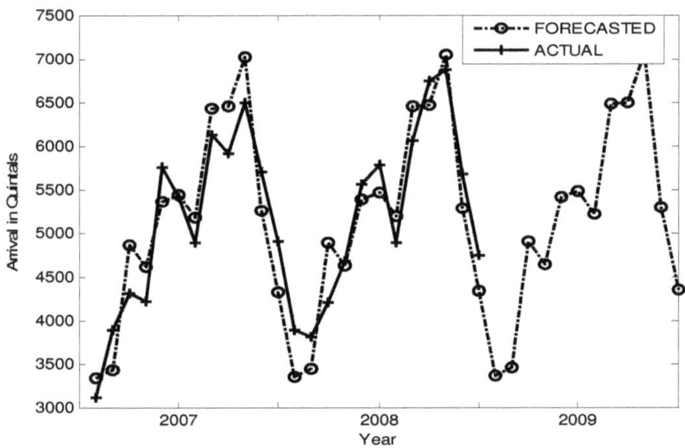

Fig 4.7 Graphical representation of actual and forecasted arrivals of banana from January 2007 onwards.

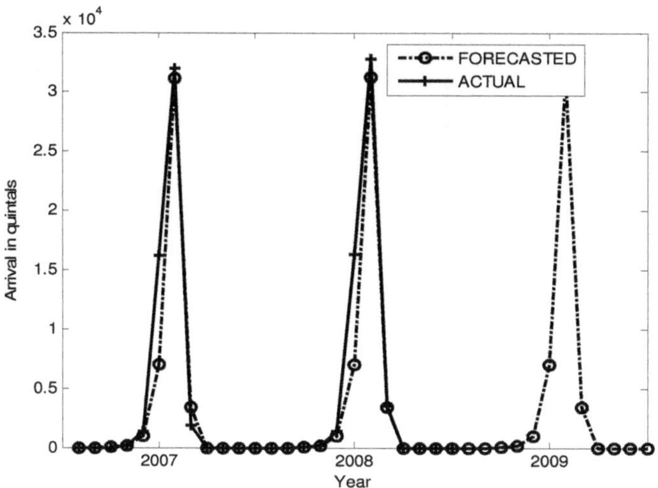

Fig 4.8 Graphical representation of actual and forecasted arrivals of mango from January 2007 onwards.

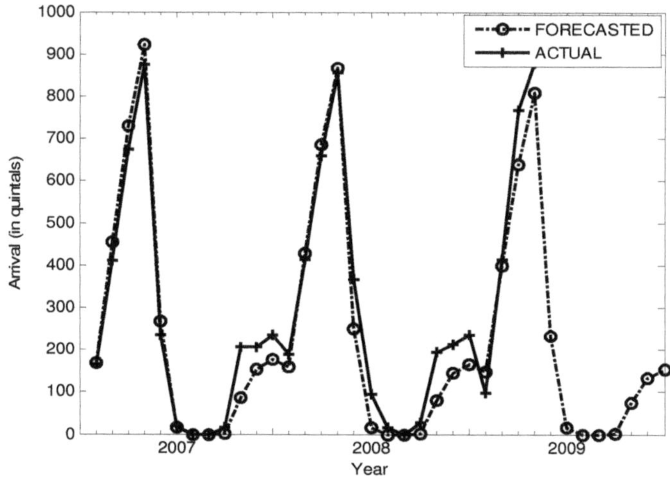

Fig 4.9 Graphical representation of actual and forecasted arrivals of orange from January 2007 onwards.

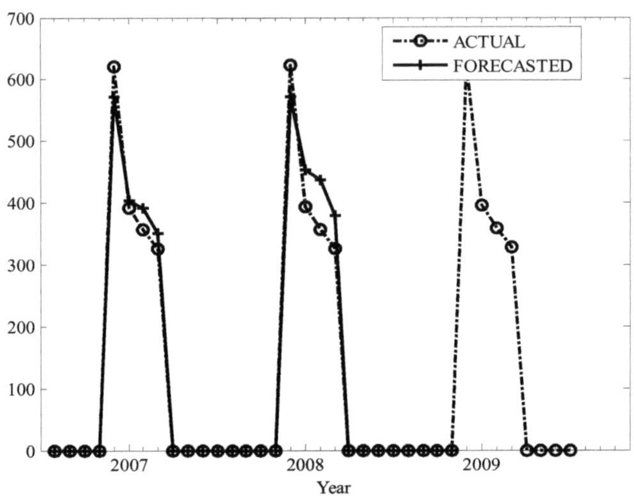

Fig 4.10 Graphical representation of actual and forecasted arrivals of peach from January 2007 onwards.

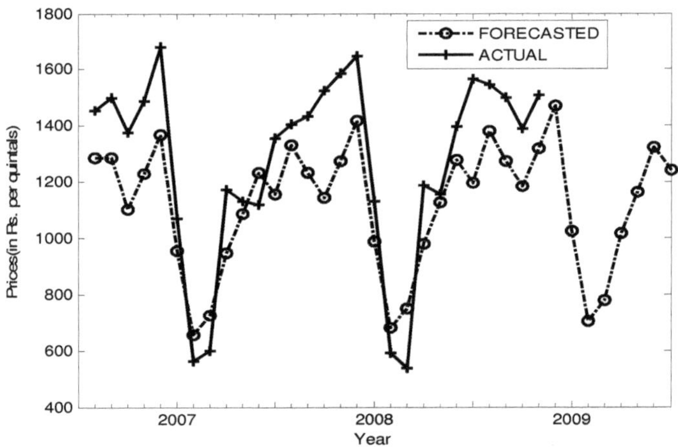

Fig 4.11 Graphical representation of actual and forecasted prices of apple from January 2007 onwards.

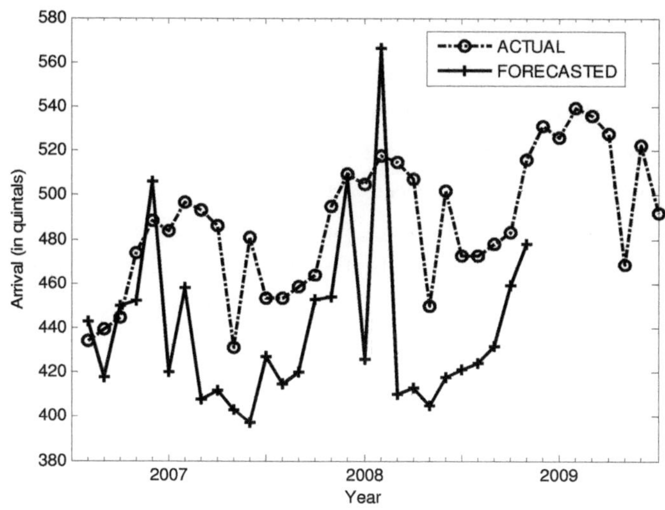

Fig 4.12 Graphical representation of actual and forecasted prices of banana from January 2007 onwards.

5. SUMMARY AND CONCLUSIONS

In this dissertation an attempt was made to analyse the fruit arrival and price patterns in a wholesale fruit market (Haldwani Mandi) for trend and seasonality in order to develop forecasting models for the fruit arrival process so as to rationalize an important input to fruit mandi system design. Historical time series data on monthly arrivals and average monthly prices was collected from the Haldwani mandi records for the period January, 1990 to April, 2009. Arrivals and prices of five fruits (Mango, Banana, Apple, Peach and Orange) were considered. A program in MATLAB 7.0 was developed for trend and seasonal analysis. For forecasting, these trend models were extended and seasonal index were applied for each month. Forecasting models were developed on the basis of first 204 months (January 1990 to December 2006) data using time series analysis technique. Forecasts were generated for the next 36 months (January 2007 to December 2009). These forecasts were compared with the actual arrivals for January 2007 to April 2009. Results of the present investigation have shown that:

1. Peak arrival of Apple was observed in the month of July, Banana arrived maximum in October, whereas Mango, Orange and Peach recorded maximum arrival in the months of July, April and May respectively. Seasonal indices corresponding to these peak arrivals were 626.66, 136.29, 876.06, 371.17 and 439.89 respectively.

2. In the case of Apple, arrivals and prices were inversely correlated; i.e., prices were low during peak arrival. However, in the case of banana, there was very little correlation; one major reason for this seems to be that the prices of Banana did not vary throughout the year.

3. There was a increasing trend in the case of prices.

4. In the trend analysis, it was observed that the arrival of Apple and Orange had a decreasing trend for the years under analysis, whereas Mango and Banana had an increasing trend for this period. The monthly arrival of Peach almost remained the same.

5. Future forecasts were developed for the monthly arrival and average monthly prices of these selected commodities. Maximum error for each forecast was calculated on the basis of peak arrival/ prices in a year and maximum error was obtained as 26.89% in forecasting the arrival of Apple.

REFERENCES

Anand, V. and Sareen, V. 2007. Probability analysis of various fruit and vegetable arrival in Haldwani Mandi. B. Tech. Thesis in Agricultural Engineering, G. B. Pant University of Agriculture and Technology, Pantnagar.

Baudentistl, W. 1971. Investigation of trends in total milk supplies and the market for dairy products in Austria for the year 1960 to 1970. Osterreichische Milchwirtschaft. 26 : 337-343 [De].

Blyn, G. 1973. Price series correlations as a measure of market integration. Indian Journal of Agricultural Economics. 28: 56-59.

Boyle, G. 1980. A time series analysis of pigs received at Irish bacon factories. Irish Journal of Agricultural Economics and Rural Sociology. 8 : 179-189.

Hanseens, D.M.I.A. 1977. An empirical study of time series analysis in marketing model building. Thesis, Doctor of Philosophy. Purdue University, Purdue. 162 p.

Jain, H.K. and Kaul, J.L. 1980. A special analysis of potato arrivals and price in Punjab, Agricultural Marketing (Nagpur). 22: 5-12.

Kalloli, M.M., Hiremath, K.C. and Rao, K.A. 1988. An economic analysis of efficiency in grading and bricking operations of groundnut in Karnataka. Indian Journal of Agricultural Economics 43(3): 488-496.

Lundahl, M. and Peterson, E. 1982. Price series correlation and market integration: some further evidence. Indian Journal of Agricultural Economics. 37: 184-190.

Narain, Maharaj. 1984. Modelling and forecasting grain Mandi arrivals and prices for system design. Ph.D. Thesis (Agricultural Engineering), G. B. Pant University of Agriculture and technology, Pantnagar.

Ngenge, A.W. 1982. A time series analysis of the efficiency of selected U.S. farm commodity markets. Dissertation Abstracts International. A 43: 2043.

Pandey, R.K. and Mishra, A. 2009. Application of artificial Neural Networks in Predicting monthly arrival and prices of various fruits at 'Haldwani Mandi'. National Conference on

Engineering for Food and Bio-Processing, 27 February-1 March 2009, College of Technology, Pantnagar.

Pouch, S. 1997. The world fruit trade: the emergence of the south. Chambers- d' Agriculture. 851: 4.

Sharma, J.S. and Roy, S. 1979. Food grain production and consumption behaviour in India, 1960-77. *In:* Two analysis of Indian Food Grain Production and Consumption data. Washington, International Food Policy Research Institute. Research Report No. 12. Pp. 11-52.

Singh, B. 1975. An analytical study of interrelationship among arrivals, prices and production of groundnut in the Punjab state. Journal of Research, Punjab Agricultural University. 12: 406-411.

Singh, B and Sindhu, D.S. 1972. Pattern of market arrivals and prices of food grains in Punjab. Agricultural Marketing (Nagpur). 14: 10-15.

Singh, B.P.N. and Gupta, D.K. 1982. Grain inflow characteristics of an Indian Mandi – a case study. *In:* International Warehousing Conference Proceedings, 3 – 6 November 1982, New Delhi.

Singh, B.; Singh, B and Hundal, L.S. 1979. Seasonal variations in production, marketed surplus and prices of milk in urban and rural areas of Punjab. *In:* 39th Annual Conference of the society of the Agricultural Economics, 18-20 December 1979, Bangalore (India).

Singh, R. and Sajwan, R. 2008. Time series analysis of selected fruits arrival and prices in Haldwani Mandi. B. Tech. Thesis in Agricultural Engineering, G. B. Pant University of Agriculture and Technology, Pantnagar.

Tijskens, L.M.M., Vollebregt, H.M., Tanner, D.J. and Fores-Sith, M.R. 2003. Prediction of storage or shelf life for cool stored fresh produce. Acta, Horticulture', 604, 305-311.

Wegner, J. 1989. International markets for fruit- analysis of supply, demand, international trade and prices using the example of selected types of fruit. Agrawirtschaft, Sonderheft **12:** 309.

Yadav, S.M., Shrivastav, S., and Singh, P. 1999. Probability analysis of fruits and vegetables arrival in Haldwani Mandi. B. Tech. Thesis in Agricultural Engineering, G. B. Pant University of Agriculture and Technology, Pantnagar.

Yuan, S.L. 1985. A study on planning fruit and vegetable wholesale markets in southern Taiwan. Journal of Agricultural Economics, Taiwan. **38:** 25-69.